Leaping Learners
Education, LLC

For more information and resources visit us at:
www.leapinglearnersed.com

Every attempt has been made to credit each photo. Please contact us if there has been an error and we will resolve the issue.

Photo Credits

Cover, © robynmac/stock.adobe.com, Pg. 5-6 © mtilghma/stock.adobe.com; Pg 7-8 © Andreas/Stock.adobe.com; Pg 9-10 © hotshotsworldwide/stock.adobe.com; Pg 11-12 © senk/ stock.adobe.com; Pg.15-16 © Cathy Keifer/stock.adobe.com; Pg. 17-18 © asbtkb/stock.adobe. com; Pg. 19-20 © ookawaphoto/stock.adobe.com; Pg. 21-22 © kikkerdirk/stock.adobe.com, © robynmac/stock.adobe.com; Pg. 23-24 © http://farm8.staticflickr.com/7167/6407253419_2 45d5eb519_o.jpg; Pg 25-26 © Jakinnboaz/stock.adobe.com; Pg 27 © https://www.xtof.photo/ porfolios/costarica/

All design by Sean Bulger
All other pictures by Sean Bulger or royalty free from Pixabay.com

ISBN

978-1-948569-16-3

Dear Parents and Guardians,

Thank you for purchasing a *Matt Learns About* series book! After teaching students from kindergarten to second grade for more than seven years, I became frustrated by the lack of engaging books my students could read independently. To help my students engage with nonfiction topics, my wife and I decided to write nonfiction books for children. We hope to inspire young children to learn about the natural world.

Here at Leaping Learners Education, LLC, we have three main goals:

1. Spark young readers' curiosity about the natural world
2. Develop critical independent reading skills at an early age
3. Develop reading comprehension skills before and after reading

We hope your child enjoys learning with Matt. If you or your children are interested in a topic we have not written about yet, send us an email with your topic, and maybe your book will be next!

Thank you,

Sean Bulger, Ed.M

www.leapinglearnersed.com

Reading Suggestions:

Before reading this book, encourage your children to do a "picture walk," where they skim through the book and look at the pictures to help them think about what they already know about the topic. Thinking about what they already know helps children get excited about learning more facts and begin reading with some confidence.

Preview any new vocabulary words with your child. Key vocabulary words are found on the last few pages of the book. Have your children use the new phrases in their own words to see if they understand the definition.

After previewing the book, encourage your children to read the book independently more than once. After they have read it, ask them specific questions related to the information in the book. Encourage them to go back and reread the relevant section in the book to retrieve the answer in case they forgot the facts.

Finally, see if your child can complete the reading comprehension exercises at the end of the book to strengthen their understanding of the topic!

This book is best for ages 6-8
but. . .
Please be mindful that reading levels are a guide and vary depending on a child's skills and needs.

Matt Learns About . . . Red-Eyed Tree Frogs

Written by Sean and Anicia Bulger

Table of Contents

Hi! My name is Matt. As you can see, I am a monkey! I love to discover and learn new things. In this book, we will learn about red-eyed tree frogs. Let's go!

Introduction

What's green, orange, and lives in rainforests?

PG 2

The red-eyed tree frog!

Habitat

Where does the red-eyed tree frog live?

The red-eyed tree frog lives in the **canopy** layer of some rainforests. It likes to live in warm, wet, and **swampy** parts of rainforests.

PG 5

Body

The red-eyed tree frog has a green body with orange toes. It's so small, it can fit in your hand!

PG 7

What does the red-eyed tree frog look like?

The red-eyed tree frog has sticky feet with suction cups. This helps the frog grab onto branches and move in the forest.

The red-eyed tree frog has BIG red eyes. These eyes help it see in the dark.

Eye

PG 12

Food

Which animals does the red-eyed tree frog eat?

Moth

The red-eyed tree frog's favorite foods are flies, spiders, grasshoppers, and moths.

Spider

Grasshopper

PG 14

The red-eyed tree frog hunts for its **prey** at night.

It has a long, sticky tongue. It uses its tongue to catch food.

Tongue

PG 16

Predators

Which animals eat the red-eyed tree frog?

Some animals eat the red-eyed tree frog. Its main **predators** are bats and snakes.

PG 18

Staying Safe

If the red-eyed tree frog feels scared, it makes its body a dark green. This can help it blend in with leaves.

To stay safe, the red-eyed tree frog makes its body flat on a leaf while it sleeps. This makes it hard for its predators to see it.

Frog Life-cycle

How does the red-eyed tree frog have babies?

Unlike most frogs that lay their eggs in water, the red-eyed tree frog lays its eggs in branches above water. This keeps them away from hungry fish.

Eggs

The mother frog covers the eggs in a thick jelly to keep them safe. When the babies **hatch**, they fall into the water below.

Jelly

A frog's life cycle keeps repeating. A frog lays eggs, then the eggs hatch and **tadpoles** come out. The tadpoles slowly grow legs and change into frogs. Then the new frog lays eggs.

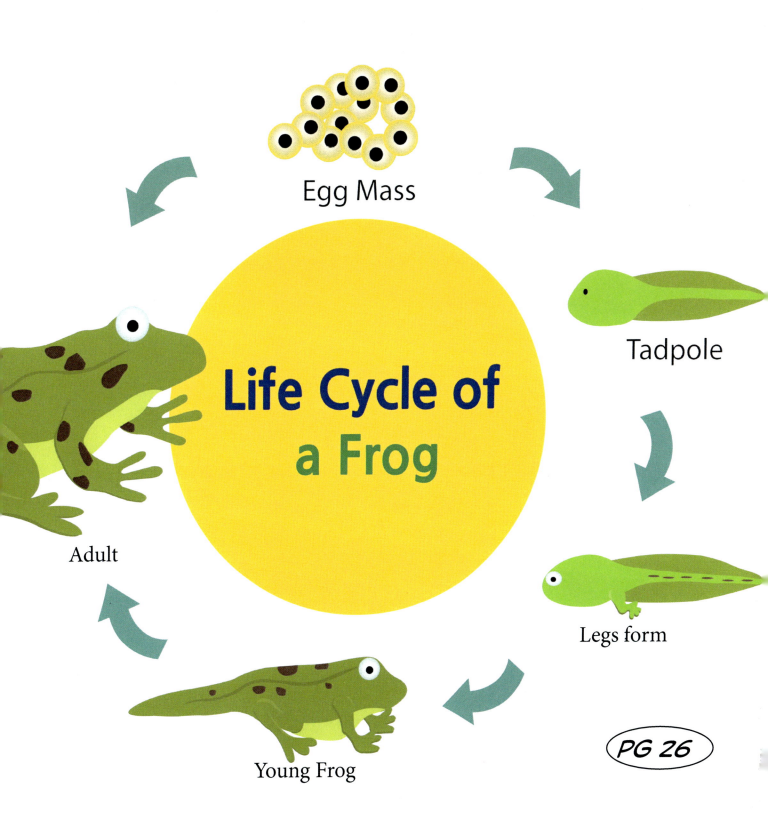

Egg Mass

Tadpole

Life Cycle of a Frog

Adult

Legs form

Young Frog

PG 26

What are some interesting facts about red-eyed tree frogs?

!!!!!

The male frog gets a piggy-back ride from the female before she lays her eggs.

Red-eye tree frogs use their eyes to help them swallow food. Their big eyes help them to push down the food.

PG 27

Glossary

A glossary tells the reader the meaning of important words.

Canopy – Second tallest layer of the rainforest

Swamp – Wet land with lots of water and mud

Prey – The food of other animals

Predator – Animal that eats other animals

Hatch – When a baby is born out of an egg

Tadpole – Baby frog with tail and gills

Draw a picture of the
red-eyed tree frog.

Draw a picture of a red-eye
tree frog eating.

The red-eye tree frog is eating a _____.

Quiz

1. Which layer of the rainforest does the red-eyed tree frog live in?

a. Canopy

b. Emergent

c. Forest floor

2. What does the red-eyed tree frog's feet help them do?

a. Catch Food

b. Stick to branches

c. Protect itself

3. What is the main idea of the section called "Food"?

a. Animals that eat red-eyed tree frogs

b. Cooking red-eyed tree frogs

c. What red-eyed tree frogs eat

Common core standards:

RI. 1. 1 - Questions 1, 2

RI. 1. 2 - Question 3

4. Why do red-eyed tree frogs lay their eggs in branches above water?

a. To keep them safe from fish

b. They like to be high up

c. Tadpoles can't swim

5. What are two ways an adult red-eyed tree frog stays safe?

a. Hops fast, makes loud noises

b. Poison skin, big teeth

c. Turns dark green, lays flat on a leaf

6. What does the diagram on page 25 teach you about?

a. The life cycle of a frog

b. Different species of frogs

c. How frogs can hop in a circle

Common core standards:

RI. 1.1 - Questions 4

RI. 1.2 - Question 5

RI. 1.8 - Question 6

Want to learn about ocean animals? Check out the "Fay Learns About..." series!

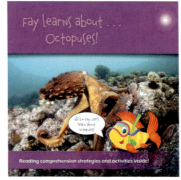

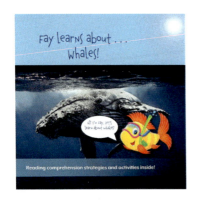

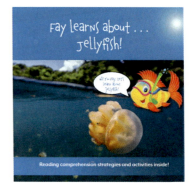

Great for emerging readers ages 6-8

Want to learn about Farm Animals? Check out the "Katie Teaches you About..." series!

Great for early readers ages 4-6

Want to learn about colors? Check out the "Clayton Teaches you About..." Series!

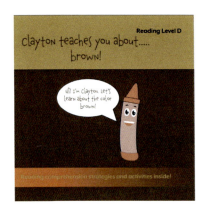
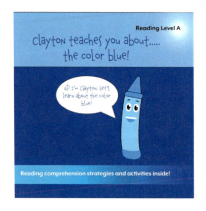
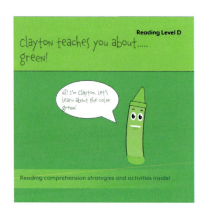

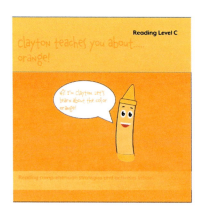
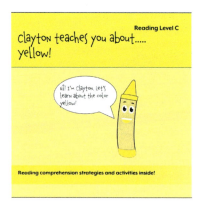

Great for early readers ages 4-6

Made in United States
Orlando, FL
12 October 2023

37807477R00024